Elementary Physics

Gravity

THOMSON
★
GALE

San Diego • Detroit • e • London • Munich

THOMSON
GALE

For more information, contact
The Gale Group, Inc.
27500 Drake Rd.
Farmington Hills, MI 48331-3535
Or you can visit our Internet site at http://www.gale.com

Photo Credits: **Art Explosion:** 3t; **Art I Need:** 4; **The Brown Reference Group plc:** 2b; **Corbis:** Morton Beebe 1, Richard Hutchings 2c; **NASA:** 2t, 3b, 6, 7, 12t, 14, 20; **Photodisc:** Photolink 8, 18; **US Dept of Defense:** 3c.

Consultant: Don Franceschetti, Ph.D., Distinguished Service Professor, Departments of Physics and Chemistry, The University of Memphis, Memphis, Tennessee

For The Brown Reference Group plc
Text: Ben Morgan
Project Editor: Tim Harris
Picture Researcher: Helen Simm
Illustrations: Darren Awuah and Mark Walker
Designer: Alison Gardner
Design Manager: Jeni Child
Managing Editor: Bridget Giles
Production Director: Alastair Gourlay
Children's Publisher: Anne O'Daly
Editorial Director: Lindsey Lowe

LIBRARY OF CONGRESS CATALOGING-IN-PUBLICATION DATA

Morgan, Ben.
Gravity / by Ben Morgan.
p. cm. — (Elementary physics)
Includes bibliographical references and index.
ISBN 1-41030-081-1 (hardback: alk. paper) — ISBN 1-41030-199-0 (paperback: alk. paper)
1. Gravity—Juvenile literature. 2. Tides—Juvenile literature. I. Title.

QC178.M67 2003
531'.14—dc21 2003002083

Printed and bound in Singapore
10 9 8 7 6 5 4 3 2 1

Contents

Gravity pulls this bungee jumper down.

What is Gravity?

If you toss a ball in the air, it soon comes down again. What makes it fall? The answer is that Earth—the **planet** we live on—pulls everything toward it. This pull, or **force**, is called **gravity**. Gravity makes leaves fall from trees. It makes streams run downhill. And it makes things fall to the ground if you drop them.

Life would be hard without gravity. You would not be able to pour a drink into a glass or run a bath. You could not lie on a bed or ride a bicycle downhill. If you tossed a ball in the air, it would keep on moving forever. You would never see it again.

Earth has a strong force of gravity.

Earth's Gravity

Earth is not the only thing that pulls other objects toward it. Your own body has **gravity**. It is very weak, though. Most small objects have such a weak **force** of gravity that you never notice it. Only really huge objects, like Earth, have gravity that is strong enough to notice. The Moon also has gravity, but it is six times weaker than Earth's gravity. **Astronauts** who landed on the Moon could jump as high as kangaroos there because of the weak gravity.

the Moon

Elephants are the heaviest land animals in the world. A full-grown male African elephant can weigh more than 100 people!

Weight

Earth's **gravity** pulls on objects by different amounts. The stronger the pull, the heavier an object feels. An elephant is heavier than a feather because gravity pulls it much more strongly. The strength of this pull is an object's weight. You can measure your own weight with a **bathroom scale**.

Earth is roughly the shape of a ball. But because it is not a perfect ball shape, its gravity is not exactly the same in all places. If you went to the South Pole, you would weigh slightly more than usual. If you climbed a mountain, you would weigh slightly less.

1

center of gravity

1 A pencil lying on its side has a wide base and a low center of gravity. It is stable.

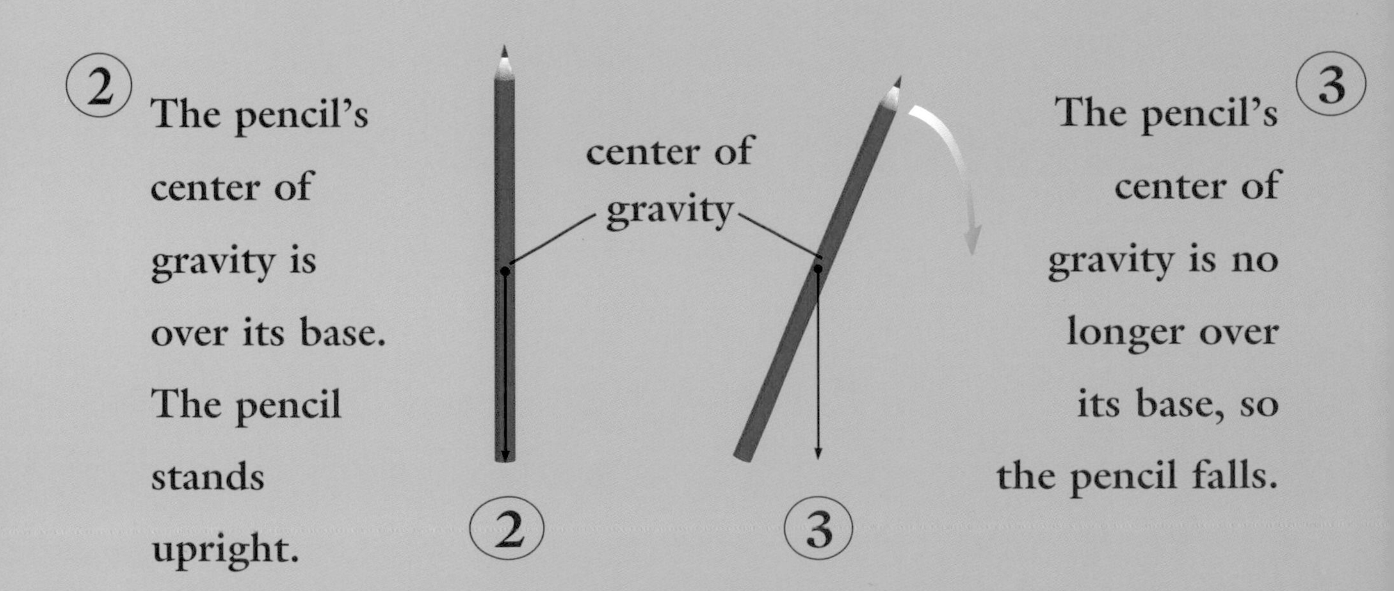

2 The pencil's center of gravity is over its base. The pencil stands upright.

3 The pencil's center of gravity is no longer over its base, so the pencil falls.

Center of Gravity

Gravity affects every part of an object. But there is one place in any object where gravity seems to act most strongly. This point is called the **center of gravity**. An object stays balanced as long as its center of gravity is over its base. If the object's center of gravity is not over its base, the object topples.

Objects that are wider than they are tall have a low center of gravity. It is more difficult to push over an object with a low center of gravity, such as a pencil on its side. This is a **stable** object. You can make a pencil **unstable** by standing it on one end. Then, you only have to blow it gently to make it fall over.

Earth moves around the Sun in a curved path called an orbit.

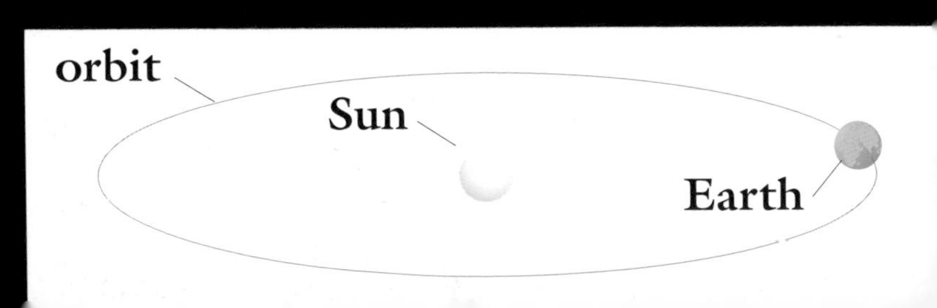

Orbits

The Sun's gravity is so enormous that it has trapped the nine **planets** of the solar system. The planets move around the Sun along curved paths. The paths are called **orbits**. The Sun's gravity keeps the planets in their orbits and stops them flying off into space.

You can see how orbits work with a simple experiment. Tie a soft toy to a piece of string. Then go outdoors. Hold the other end of the string and swing the toy around. The toy seems to try to fly away, but the string pulls it back. So the toy flies around you in an orbit.

The Moon's gravity pulls at Earth.

Tides

Although you cannot feel the Moon's **gravity**, it does affect our **planet**. The Moon's gravity causes **tides**. Seas rise and fall regularly all over Earth because the Moon tugs the seawater slightly toward it.

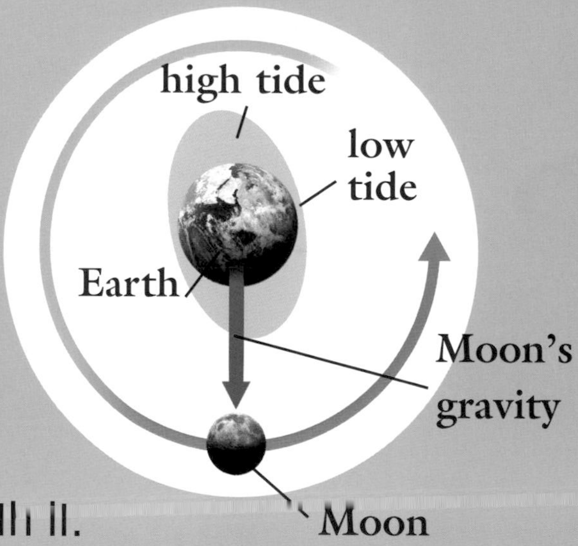

The Moon travels around Earth in an **orbit**. As the Moon moves, the high tide moves with it.

The Sun also affects tides. When the Sun and Moon are in a line, their gravity joins together to make a high tide that is even higher than usual. It is called a spring tide.

A ping-pong ball (left) is light, but it falls at the same speed as a pool ball (right).

Falling

Which do you think falls faster, a ping-pong ball or a pool ball? You might think that heavy objects fall fastest. In fact, the weight of objects makes very little difference. Drop a ping-pong ball and a pool ball at the same time. See if one reaches the ground before the other.

An Italian scientist called Galileo was the first person to prove that all objects fall at the same speed. More than 400 years ago, he dropped two weights from a tower. One weighed 10 pounds (4.5 kg). The other weighed 1 pound (0.45 kg). The two weights hit the ground at exactly the same time.

Fluffy dandelion seeds catch lots of air when they fall. The resistance of the air, or drag, slows their fall to the ground.

Defying Gravity

The fluffier or wider an object is, the more air it catches. The more air it catches, the slower it falls. This effect is called **drag**, or **air resistance**. Air resistance is what makes **parachutes** fall so gently.

Because of drag, Galileo's experiment would not have worked if he had used a feather and a metal weight. On the Moon, though, where there is no air, the experiment would have worked perfectly even with a feather. **Astronauts** who visited the Moon in the 1970s dropped a feather and a metal weight. Both objects hit the ground at the same time.

Balance Trick

center of gravity

width of base

Try this experiment to see how important an object's **center of gravity** is. First, try to make a potato balance on the rim of a plastic glass. You will probably find it is impossible. Next, stick two forks in the potato as shown. Now try again. The potato balances. How does the trick work? When the potato was on its own, it had a high center of gravity and a narrow base. That made it **unstable**, so it fell off the rim. When you stuck the forks in, you lowered the center of gravity. It was then below the potato. Also, the base of the forks was wider than the base of the potato. This made the object much more **stable**.

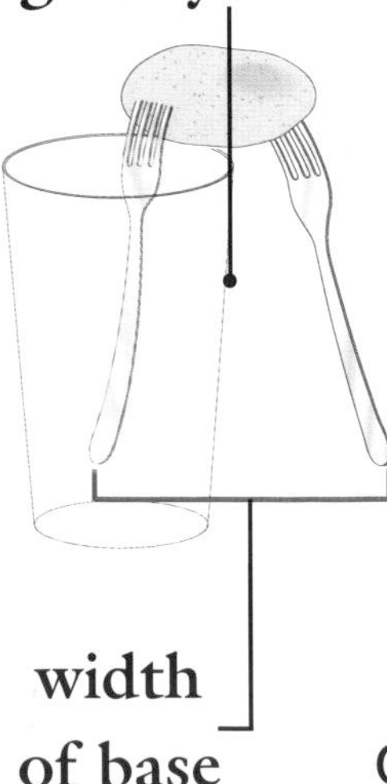

Glossary

astronaut a person who travels into outer space.

bathroom scale a machine that weighs people.

center of gravity the point in an object where gravity seems to act most strongly.

force a push or pull.

orbit the curved path that a planet follows around the Sun or that the Moon follows around Earth.

parachute a structure that slows an object as it falls through air.

planet a large body that revolves around the Sun.

stable when an object does not topple over after being pushed.

tides the regular movement of seas as they rise and fall. Tides are caused by the Moon's gravity.

unstable when an object topples over easily after being pushed.

Look Further

To find out more experiments you can carry out to show gravity, read *101 Great Science Experiments* by Neil Ardley (DK Publishing).

You can also find out more about gravity from the internet at this website: www.chem4kids.com/

Index